~A BINGO BOOK~

Geometry
Basics
Bingo Book

COMPLETE BINGO GAME IN A BOOK

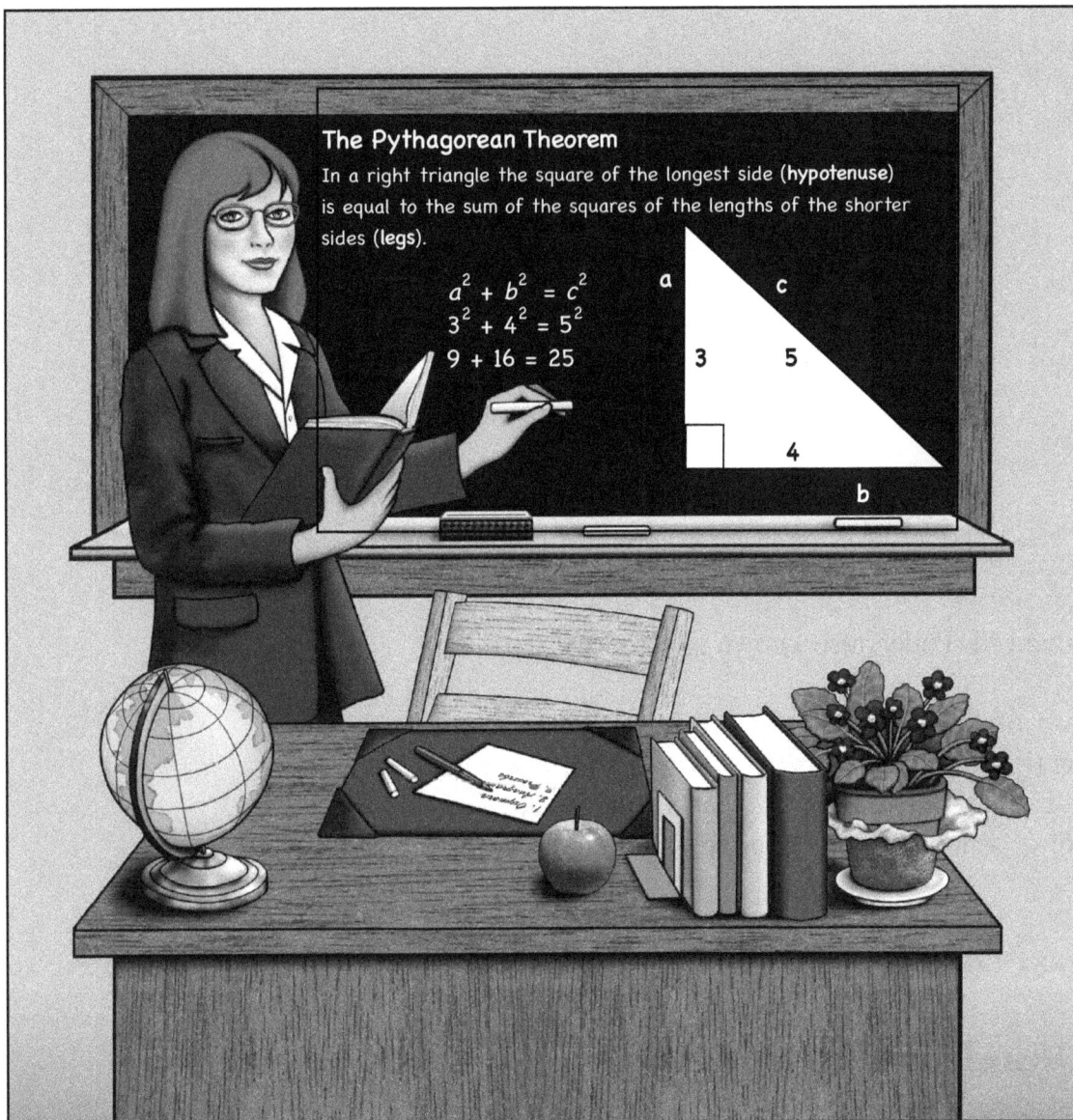

Written By Rebecca Stark

TITLE: Geometry Basics Bingo
AUTHOR: Rebecca Stark

ISBN 978-0-87386-460-2

Educational Books 'n' Bingo

Printed in the U.S.A.

GEOMETRY BASICS BINGO DIRECTIONS

INCLUDED:

List of Terms

Templates for Additional Terms and Clues

2 Clues per Term

30 Unique Bingo Cards

Markers

1. **Either cut apart the book or make copies of ALL the sheets. You might want to make an extra copy of the clue sheets to use for introduction and review. Keep the sheets in an envelope for easy reuse.**

2. Cut apart the call cards with terms and clues.

3. Pass out one bingo card per student. There are enough for a class of 30.

4. Pass out markers. You may cut apart the markers included in this book or use any other small items of your choice.

5. Decide whether or not you will require the entire card to be filled. Requiring the entire card to be filled provides a better review. However, if you have a short time to fill, you may prefer to have them do the just the border or some other format. Tell the class before you begin what is required.

6. There are 50 terms. Read the list before you begin. If there are any terms that have not been covered in class, you may want to read to the students the term and clues before you begin.

7. There is a blank space in the middle of each card. You can instruct the students to use it as a free space or you can write in answers to cover terms not included. Of course, in this case you would create your own clues. (Templates provided.)

8. Shuffle the cards and place them in a pile. Two or three clues are provided for each term. If you plan to play the game with the same group more than once, you might want to choose a different clue for each game. If not, you may choose to use more than one clue.

9. Be sure to keep the cards you have used for the present game in a separate pile. When a student calls, "Bingo," he or she will have to verify that the correct answers are on his or her card AND that the markers were placed in response to the proper questions. Pull out the cards that are on the student's card keeping them in the order they were used in the game. Read each clue as it was given and ask the student to identify the correct answer from his or her card.

10. If the student has the correct answers on the card AND has shown that they were marked in response to the *correct questions,* then that student is the winner and the game is over. If the student does not have the correct answers on the card OR he or she marked the answers in response to *the wrong questions,* then the game continues until there is a proper winner.

11. If you want to play again, reshuffle the cards and begin again.

Have fun!

TERMS/ANSWERS

6	LINE
8 CUBIC IN.	OBTUSE ANGLE
12	PARALLEL
14	PARALLELOGRAM
25 SQUARE INCHES	PERIMETER
40°	PERPENDICULAR
47°	PI (π)
180°	PLANE
360°	POINT
ACUTE ANGLE	POLYGON
ACUTE TRIANGLE	PROPORTION
ANGLE	PROTRACTOR
AREA	PYRAMID
CHORD	PYTHAGOREAN THEOREM
CIRCLE	QUADRILATERAL
CIRCUMFERENCE	RADIUS
COMPLEMENTARY	RAY
CONGRUENT	RIGHT ANGLE
CUBE	RIGHT TRIANGLE
DIAMETER	SCALENE TRIANGLE
EQUILATERAL TRIANGLE	STRAIGHT ANGLE
GEOMETRY	SUPPLEMENTARY
HYPOTENUSE	TRAPEZOID
INTERSECT	TRIANGLE
ISOSCELES TRIANGLE	VOLUME

Additional Terms

Choose as many additional terms as you would like and write them in the squares.
Repeat each as desired.
Cut out the squares and randomly distribute them to the class.
Instruct the students to place their square on the center space of their card.

Clues for
Additional Terms

Write three clues for each of your terms.

_____

1.

2.

3.

_____

1.

2.

3.

_____

1.

2.

3.

_____

1.

2.

3.

_____

1.

2.

3.

_____

1.

2.

3.

© **Barbara M. Peller**

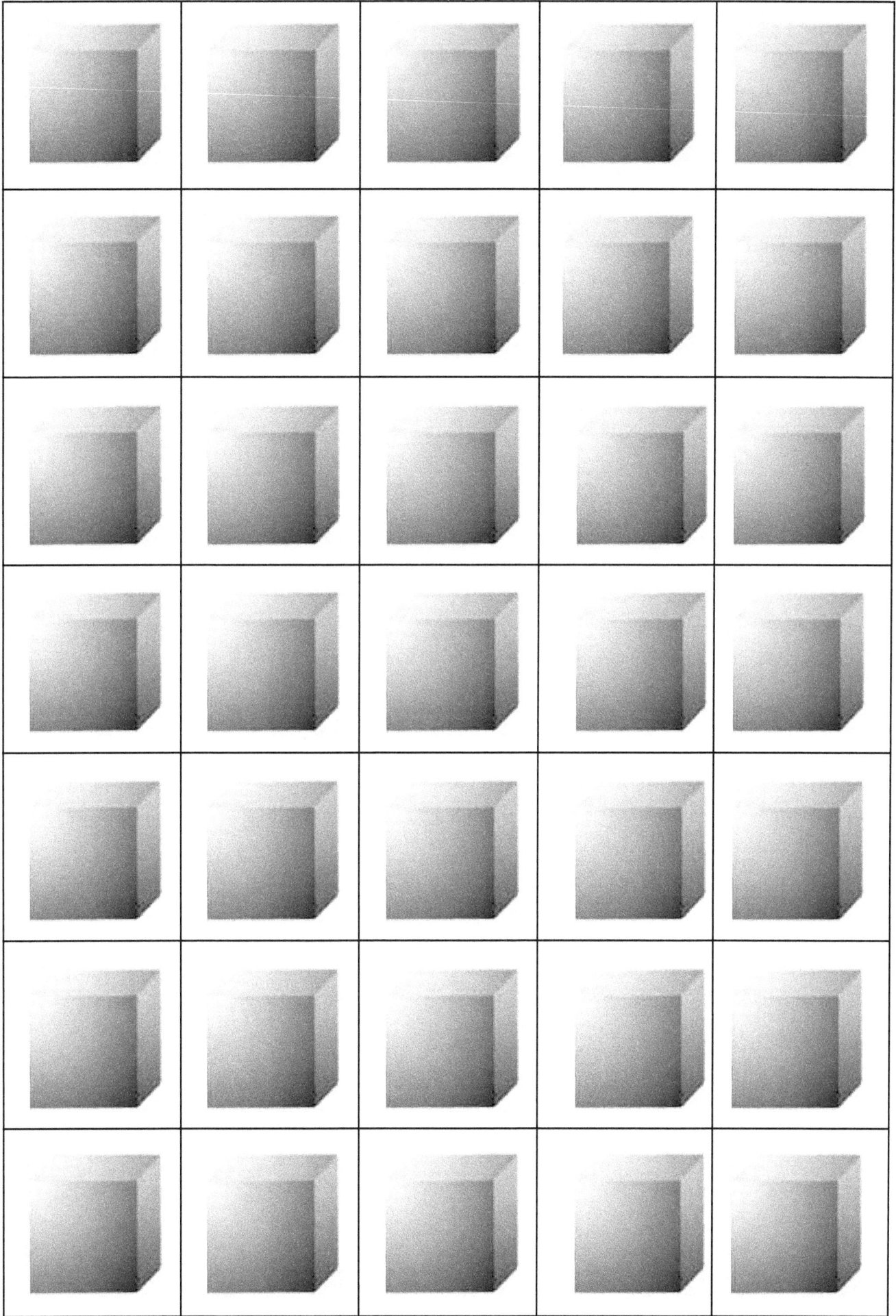

6 1. The value of y in this proportion: $$2/3 = 4/y$$ 2. $\sqrt{36}$ 3. In a right triangle, if side *b* is 8 inches, and side *c* is 10 inches, then side *a* is ___ inches. *For #3: $a^2 + b^2 = c^2$	**8 Cubic In.** 1. Volume of a cube with 2-inch sides. 2. Volume of a pyramid with a height of 6 inches and a triangular base with an area of 4 square inches.* 3. 2 in.^{3} + 6 in.^{3} = *For #2: *A = 1/3 x (base area) x height*
12 1. The area of a rectangle 3 feet by 4 feet is __ square feet. 2. If one side of an equilateral triangle measures 4 inches, its perimeter is ___ inches. 3. $\sqrt{36}$	**14** 1. $\sqrt{196}$ 2. Perimeter of a parallelogram with 2 sides that are 2 inches and 2 sides that are 5 inches is ___ inches. 3. Perimeter of an isosceles triangle with a base of 2 centimeters and a side that is 6 centimeters is ___ centimeters.
25 Square Inches 1. Area of a square whose side is 5 inches. 2. Area of a trapezoid that is 4 inches in height with bases that are 5 inches and 7.5 inches.* 3. If the radius of a circle is 5 inches, multiply π by this to find the area. *For #2: *A = height x (base1 + base2) / 2*	**40°** 1. This angle and a 140° angle are supplementary. 2. This angle and a 5° angle are complementary. 3. If Angle *A* and Angle *D* are corresponding angles and if Angle *A* is 40°, then Angle *D* is ___.
47° 1. If angles 1 and 4 are vertical angles, and angle 1 = 47°, then angle 4 = ___. 2. Its supplementary angle is 133°. 3. Its complementary angle is 43°.	**180°** 1. If two angles combine to measure ___, they are said to be supplementary angles. 2. A semicircular protractor is divided into degrees from 0° to ___. 3. The angle formed by drawing the diameter of a circle is ___.
360° 1. The sum of the interior angles of a parallelogram. 2. The sum of the interior angles of a trapezoid. 3. The sum of the interior angles of a square.	**Acute Angle** 1. It measures less than 90°. 2. It is smaller than a right angle. 3. A 60° angle is one.

Geometry Basics Bingo

Acute Triangle 1. This polygon has 3 acute angles. 2. If all 3 of a triangle's angles are unequal and less than 90°, it is this kind of triangle. 3. An equilateral triangle is also this kind of triangle.	**Angle** 1. One is formed when 2 rays have a common endpoint. 2. Its common endpoint is called a vertex. 3. We measure it in degrees.
Area 1. The measure of square units within a figure. 2. You get the ___ of a rectangle by multiplying its length times its width. 3. You get the ___ of a triangle by using the formula *A = 1/2 bh,* where *b* is the base and h is the height.	**Chord** 1. Any line segment that joins two points on a circle. 2. A diameter is one, but a radius is not. 3. If a ___ goes through the center point of a circle, we call it a diameter.
Circle 1. A closed curve made up of all the points at the same distance from a given point is one. 2. A v-shaped instrument called a compass is sometimes used to draw one. 3. The center of this closed curve is a point, which can be used to name the figure.	**Circumference** 1. The distance around a circle. 2. The name for the perimeter of a circle. 3. To find it, multiply π times the diameter.
Complementary 1. A pair of angles that equal 90° when combined are said to be this. 2. A 75° angle and a 15° angle are this. 3. A 60° angle and a 30° angle are this.	**Congruent** 1. Figures with the same size and shape are this. 2. Corresponding angles are always ___. 3. Alternate and vertical angles are always ___.
Cube 1. This figure has 6 congruent faces. 2. All 6 of this figure's faces are square. 3. If each of its sides is 3 inches, its volume is 27 cubic inches.	**Diameter** 1. A chord that passes through the center of a circle. 2. It equals two radii. 3. To find the circumference of a circle, we multiply this times π.

Geometry Basics Bingo

Equilateral Triangle 1. This figure has 3 equal sides and 3 equal angles. 2. Each of its 3 angles measures 60°. 3. A triangle with three 3-inch sides is this.	**Geometry** 1. Branch of mathematics that deals with points, lines, angles and surfaces. 2. Branch of mathematics concerned with the size, shape and relative positions of figures. 3. Study of 2- and 3-dimensional figures.
Hypotenuse 1. The longest side of a right triangle. 2. The side opposite the right angle in a right triangle. 3. In the formula $a^2 + b^2 = c^2$, c represents the ___ of a right triangle.	**Intersect** 1. To cross each other at a point. 2. What perpendicular lines do at right angles. 3. What a pair of parallel lines never do.
Isosceles Triangle 1. This kind of triangle has 2 equal sides and 2 equal angles. 2. A triangle with two 65° angles is one. 3. A triangle with the angles 40°, 40° and 100° is an obtuse ___ triangle.	**Line** 1. A set of points along a straight path. 2. A ___ segment includes all the points that fall between two endpoints. 3. Arrows pointing in both directions written over 2 points signifies a ___.
Obtuse Angle 1. It measures more than 90° but less than 180°. 2. A 120° angle is one. 3. A 175° angle is one.	**Parallel** 1. Lines that are in the same plane but never meet are called ___ lines. 2. ___ lines are at the same distance apart at all points. 3. These lines never intersect.
Parallelogram 1. A quadrilateral whose opposite sides are parallel is one. 2. To find its area, use the formula $A = bh$ (Area + base x height). 3. A square is one; so is a rhombus.	**Perimeter** 1. The distance around a geometric shape. 2. The ___ of a polygon is found by adding all of its sides. 3. If one side of a square is 3 inches, its ___ is 12 inches.

Geometry Basics Bingo

Perpendicular	**Pi (π)**
1. Lines that intersect at right angles are called ___ lines. 2. When ___ lines intersect, they form square corners at the point where they meet. 3. Four right angles are formed by the intersection of lines of this type.	1. It represents the ratio of the circumference of a circle to its diameter; its value is about 3.1459. 2. To find the circumference of a circle, multiply this times the diameter. 3. To find the area of a cylinder, use this formula: 2(___r^2) + (2___r) x h. (r = radius and h = height).
Plane	**Point**
1. A 2-dimensional surface. 2. The set of all the points on an endless flat surface. 3. A flat surface that extends into infinity in all directions.	1. An exact location in space. 2. In the line segment BC, B is an end___; so is C. 3. In the ray BC, B is an end___ and C is a ___ on the line.
Polygon	**Proportion**
1. A closed figure made up of 3 or more line segments and the same number of angles. 2. A hexagon is one with 6 sides. 3. An octagon is one with 8 sides.	1. An equation with a ratio on either side of the equal sign. 2. It is a mathematical statement that says two ratios are equal. 3. If 1 of the 4 numbers in a ___ is a variable, use cross products to find the value of the variable.
Protractor	**Pyramid**
1. An instrument used to measure and draw angles. 2. These instruments are either circular or semicircular. 3. A semicircular one is divided into degrees from 0° to 180°.	1. The base of a square ___ is a square, and all its triangular faces are congruent isosceles triangles. 2. To find the surface area of a regular___, use this formula: *1/2 x perimeter x side length* 3. A ___ is made by connecting a base to an apex.
Pythagorean Theorem	**Quadrilateral**
1. It states that the square of the hypotenuse of a right triangle is equal to the sum of the squares on the other two sides. 2. This theorem is named after a Greek mathematician named Pythagoras. 3. It can be expressed as $a^2 + b^2 = c^2$.	1. Any 4-sided polygon. 2. A parallelogram is this type of polygon; so is a trapezoid. 3. The sum of its 4 interior angles is 360°.

Geometry Basics Bingo

Radius 1. Any line segment with one endpoint at the center of the circle and the other on the circle. 2. It equals one-half of the diameter. 3. A diameter is really two of these.	**Ray** 1. A segment of a line that extends from 1 endpoint and continues indefinitely in the other direction. 2. You name it by is endpoint and a point on the line. 3. The letters *BC* under an arrow pointing right is called ___ *BC*.
Right Angle 1. An angle that measures 90°. 2. It is larger than an acute angle and smaller than an obtuse angle. 3. This kind of angle forms a square corner.	**Right Triangle** 1. This kind of triangle has a right angle. 2. Two of its 3 angles add up to 90°. 3. It has 3 sides and 3 angles; one of its angles is 90°.
Scalene Triangle 1. If all 3 sides of a triangle are a different length, it is this kind of triangle. 2. A triangle with these lengths is one: 3 cm, 4 cm and 5 cm. 3. A triangle with 2 equal sides is not one.	**Straight Angle** 1. It measures 180°. 2. It is larger than an obtuse angle. 3. It twice as large as a right angle.
Supplementary 1. A pair of angles that equal 180° when combined are said to be this. 2. A 160° angle and a 20° angle are this. 3. A 125° angle and a 55° angle are this.	**Trapezoid** 1. A quadrilateral with only 2 parallel sides. 2. It is like a parallelogram in that it has 4 sides. It is unlike a parallelogram because only 2 of its sides are parallel. 3. To find the area of this kind of quadrilateral, use the formula $A = 1/2\ (b_1 + b_2)$. (*b* refers to base)
Triangle 1. This polygon has 3 sides and 3 angles. 2. Its 3 interior angles add up to 180°. 3. To name it, use its 3 vertices.	**Volume** 1. The measure of the amount of space in a container. 2. To find the ___ of a cube, use the formula $V = s^3$ (side cubed). 3. Multiply length *x* width *x* height to find the ___ of a rectangular 3-dimensional figure.

Geometry Basics Bingo

Geometry Basics Bingo

Pythagorean Theorem	Acute Angle	6	47°	Parallelogram
Equilateral Triangle	Diameter	Scalene Triangle	40°	Triangle
Area	Straight Angle		Hypotenuse	Perimeter
8 Cubic In.	14	Volume	Polygon	Obtuse Angle
Ray	Intersect	Complementary	Pi (π)	Perpendicular

Geometry Basics Bingo

8 Cubic In.	Area	Point	Trapezoid	Fahrenheit
Obtuse Angle	40°	Circle	14	Radius
Cube	Intersect		Geometry	Volume
360°	Right Triangle	Straight Angle	Right Angle	Perpendicular
Triangle	Scalene Triangle	Complementary	Equilateral Triangle	Pi (π)

Geometry Basics Bingo: Card No. 2

Geometry Basics Bingo

8 Cubic In.	Volume	40°	Polygon	Area
Intersect	Acute Triangle	Congruent	Acute Angle	180°
14	Scalene Triangle		Radius	Angle
Straight Angle	Cube	Ray	360°	Point
Pi (π)	Equilateral Triangle	Complementary	Right Angle	Geometry

Geometry Basics Bingo

Straight Angle	Radius	6	Equilateral Triangle	Geometry
Plane	Chord	Acute Angle	Trapezoid	Area
Hypotenuse	360°		Parallelogram	47°
Volume	24 Sq. Inches	Scalene Triangle	Complementary	Circle
180°	Triangle	Proportion	Pi (π)	Perimeter

Geometry Basics Bingo

Triangle	Parallelogram	14	Circle	Equilateral Triangle
Plane	Volume	Congruent	Fahrenheit	Acute Triangle
6	Perimeter		Protractor	Line
Perpendicular	Geometry	Pythagorean Theorem	Right Angle	Diameter
40°	Complementary	Area	Straight Angle	Hypotenuse

Geometry Basics Bingo: Card No. 5

Geometry Basics Bingo

Angle	Radius	Point	Geometry	Perimeter
Polygon	14	Diameter	Acute Angle	Area
Trapezoid	Parallel		Chord	Fahrenheit
Complementary	Ray	Right Angle	Proportion	6
Obtuse Angle	Volume	Pythagorean Theorem	Hypotenuse	24 Sq. Inches

Geometry Basics Bingo

Pythagorean Theorem	Radius	Line	Protractor	40°
Parallel	Fahrenheit	Intersect	Circumference	Plane
Point	47°		Geometry	Chord
Straight Angle	360°	Congruent	8 Cubic In.	Cube
Complementary	Equilateral Triangle	Right Angle	Proportion	Angle

Geometry Basics Bingo

Hypotenuse	Radius	Circumference	Polygon	Chord
Plane	6	Trapezoid	Perimeter	Circle
24 Sq. Inches	Pyramid		Geometry	Parallelogram
Pi (π)	Straight Angle	8 Cubic In.	Parallel	360°
Scalene Triangle	Complementary	Proportion	14	Obtuse Angle

Geometry Basics Bingo

Geometry	40°	Intersect	24 Sq. Inches	Equilateral Triangle
Parallel	Fahrenheit	Hypotenuse	14	Radius
180°	Pythagorean Theorem		Diameter	Acute Triangle
Acute Triangle	Perpendicular	Ray	Protractor	Line
360°	Right Angle	Congruent	8 Cubic In.	Parallelogram

Geometry Basics Bingo

8 Cubic In.	Polygon	Chord	Trapezoid	24 Sq. Inches
Perimeter	Circle	Acute Angle	Diameter	Geometry
Pyramid	Radius		47°	Cube
Ray	Perpendicular	Acute Triangle	Right Angle	180°
Congruent	Parallel	Point	Triangle	Hypotenuse

Geometry Basics Bingo

Angle	Radius	14	Diameter	Parallel
Circumference	180°	Protractor	Fahrenheit	Acute Angle
Plane	Geometry		Point	Intersect
Congruent	Area	Right Angle	Equilateral Triangle	8 Cubic In.
Obtuse Angle	Complementary	Pythagorean Theorem	Proportion	40°

Geometry Basics Bingo

40°	Parallelogram	Complementary	Polygon	Geometry
Intersect	Obtuse Angle	6	Proportion	Diameter
Pythagorean Theorem	Line		Perimeter	Trapezoid
180°	360°	Fahrenheit	8 Cubic In.	Plane
Radius	Circumference	Pyramid	Parallel	Circle

Geometry Basics Bingo: Card No. 12

Geometry Basics Bingo

Diameter	Parallelogram	Angle	Straight Angle	Perimeter
6	Circumference	Fahrenheit	Geometry	Cube
Polygon	Circle		Intersect	Line
Hypotenuse	Right Angle	Chord	Pyramid	8 Cubic In.
Complementary	Perpendicular	Proportion	Pythagorean Theorem	Protractor

Geometry Basics Bingo

Equilateral Triangle	Fahrenheit	14	Geometry	Obtuse Angle
Circle	Pythagorean Theorem	180°	Diameter	Radius
Volume	47°		Point	Congruent
Perpendicular	Right Angle	Pyramid	Chord	Angle
Complementary	Trapezoid	Cube	Parallel	Hypotenuse

Geometry Basics Bingo

Protractor	Fahrenheit	14	40°	Polygon
Angle	Point	Acute Angle	6	Obtuse Angle
Perimeter	Pythagorean Theorem		Area	Radius
Complementary	180°	Circumference	Right Angle	Diameter
Parallel	360°	Proportion	24 Sq. Inches	Intersect

Geometry Basics Bingo

Chord	180°	Circumference	24 Sq. Inches	Right Triangle
Trapezoid	Cube	Line	Plane	47°
Acute Triangle	Parallelogram		Perimeter	Intersect
Straight Angle	Circle	Complementary	Protractor	8 Cubic In.
Parallel	Supplementary	Proportion	360°	Radius

© Barbara M. Peller

Geometry Basics Bingo

Congruent	Quadrilateral	Isosceles Triangle	180°	Equilateral Triangle
Protractor	Parallel	Right Angle	47°	Line
Fahrenheit	Hypotenuse		Supplementary	Circumference
Perpendicular	Obtuse Angle	8 Cubic In.	14	Cube
Ray	Acute Triangle	40°	Polygon	Parallelogram

Geometry Basics Bingo

24 Sq. Inches	Pyramid	Circle	Diameter	Trapezoid
Radius	Congruent	Ray	Perimeter	Parallel
Fahrenheit	Cube		Isosceles Triangle	6
Perpendicular	Acute Angle	Right Angle	8 Cubic In.	Point
Supplementary	180°	14	Quadrilateral	Angle

Geometry Basics Bingo: Card No. 18

Geometry Basics Bingo

Perimeter	Angle	180°	Circumference	Pyramid
Protractor	Polygon	Radius	40°	47°
Quadrilateral	Equilateral Triangle		Diameter	Area
Point	Supplementary	Ray	360°	Isosceles Triangle
6	Right Triangle	Parallel	Hypotenuse	Proportion

Geometry Basics Bingo: Card No. 19

Geometry Basics Bingo

Pyramid	Quadrilateral	Polygon	180°	Proportion
Circle	Intersect	Plane	Ray	Trapezoid
Parallelogram	Line		Straight Angle	Acute Angle
Triangle	Scalene Triangle	Pi (π)	360°	Supplementary
Volume	Hypotenuse	Right Triangle	8 Cubic In.	Isosceles Triangle

Geometry Basics Bingo: Card No. 20

Geometry Basics Bingo

Protractor	Angle	Plane	180°	Triangle
Parallelogram	Isosceles Triangle	Chord	Circumference	Pythagorean Theorem
Cube	Parallel		Quadrilateral	14
Ray	40°	Supplementary	Perpendicular	Hypotenuse
Straight Angle	Right Triangle	Proportion	Congruent	360°

Geometry Basics Bingo: Card No. 21

Geometry Basics Bingo

24 Sq. Inches	Point	Isosceles Triangle	6	Diameter
Trapezoid	Polygon	Area	Circumference	Acute Triangle
Circle	47°		Pythagorean Theorem	Line
Supplementary	Perpendicular	360°	Acute Angle	Equilateral Triangle
Right Triangle	Congruent	Quadrilateral	Cube	Plane

Geometry Basics Bingo

Chord	Quadrilateral	40°	6	Proportion
Angle	Pyramid	Parallel	Protractor	Acute Angle
Point	Diameter		Pi (π)	Pythagorean Theorem
Cube	Right Triangle	Supplementary	Congruent	360°
Triangle	Scalene Triangle	Hypotenuse	Ray	Isosceles Triangle

Geometry Basics Bingo

Chord	Pyramid	Equilateral Triangle	Quadrilateral	Circumference
Isosceles Triangle	Proportion	Plane	Trapezoid	Pythagorean Theorem
Line	24 Sq. Inches		Diameter	Cube
Triangle	Pi (π)	Supplementary	Congruent	Parallelogram
Volume	Straight Angle	Right Triangle	Polygon	Scalene Triangle

Geometry Basics Bingo: Card No. 24

Geometry Basics Bingo

Straight Angle	Plane	Quadrilateral	14	Isosceles Triangle
Acute Angle	Perpendicular	Protractor	Chord	Acute Triangle
Parallelogram	Area		Pi (π)	Supplementary
Circumference	Triangle	Scalene Triangle	Right Triangle	47°
Proportion	Equilateral Triangle	Circle	Parallel	Volume

Geometry Basics Bingo

Isosceles Triangle	Quadrilateral	Point	Trapezoid	24 Sq. Inches
Ray	Polygon	Circumference	Pyramid	Chord
Perpendicular	Pi (π)		47°	Straight Angle
Congruent	6	Triangle	Right Triangle	Supplementary
Line	Parallel	14	Scalene Triangle	Volume

Geometry Basics Bingo

Point	Circle	Quadrilateral	Pyramid	Intersect
Triangle	Pi (π)	Protractor	Supplementary	Acute Triangle
Right Angle	Scalene Triangle		Right Triangle	Straight Angle
24 Sq. Inches	Angle	Plane	Volume	Acute Angle
Parallel	47°	Isosceles Triangle	Area	Line

Geometry Basics Bingo

Perimeter	Pyramid	Area	Quadrilateral	Chord
Intersect	Isosceles Triangle	Pi (π)	Trapezoid	47°
Scalene Triangle	Cube		Line	Ray
8 Cubic In.	24 Sq. Inches	Parallel	Right Triangle	Supplementary
6	Fahrenheit	Obtuse Angle	Volume	Triangle

Geometry Basics Bingo

Isosceles Triangle	Pyramid	24 Sq. Inches	Protractor	14
Perpendicular	Ray	Plane	Line	Area
Parallelogram	Pi (π)		Perimeter	Quadrilateral
Intersect	Triangle	Geometry	Right Triangle	Supplementary
Chord	Acute Triangle	Volume	Angle	Scalene Triangle

Geometry Basics Bingo

Equilateral Triangle	Quadrilateral	Trapezoid	Geometry	Supplementary
Acute Angle	Pyramid	Point	47°	Diameter
Perpendicular	Acute Triangle		Line	Plane
Volume	Angle	6	Right Triangle	Pi (π)
Triangle	40°	Scalene Triangle	Isosceles Triangle	Area

www.ingramcontent.com/pod-product-compliance
Lightning Source LLC
Chambersburg PA
CBHW051419200326
41520CB00023B/7294

9 780873 864602